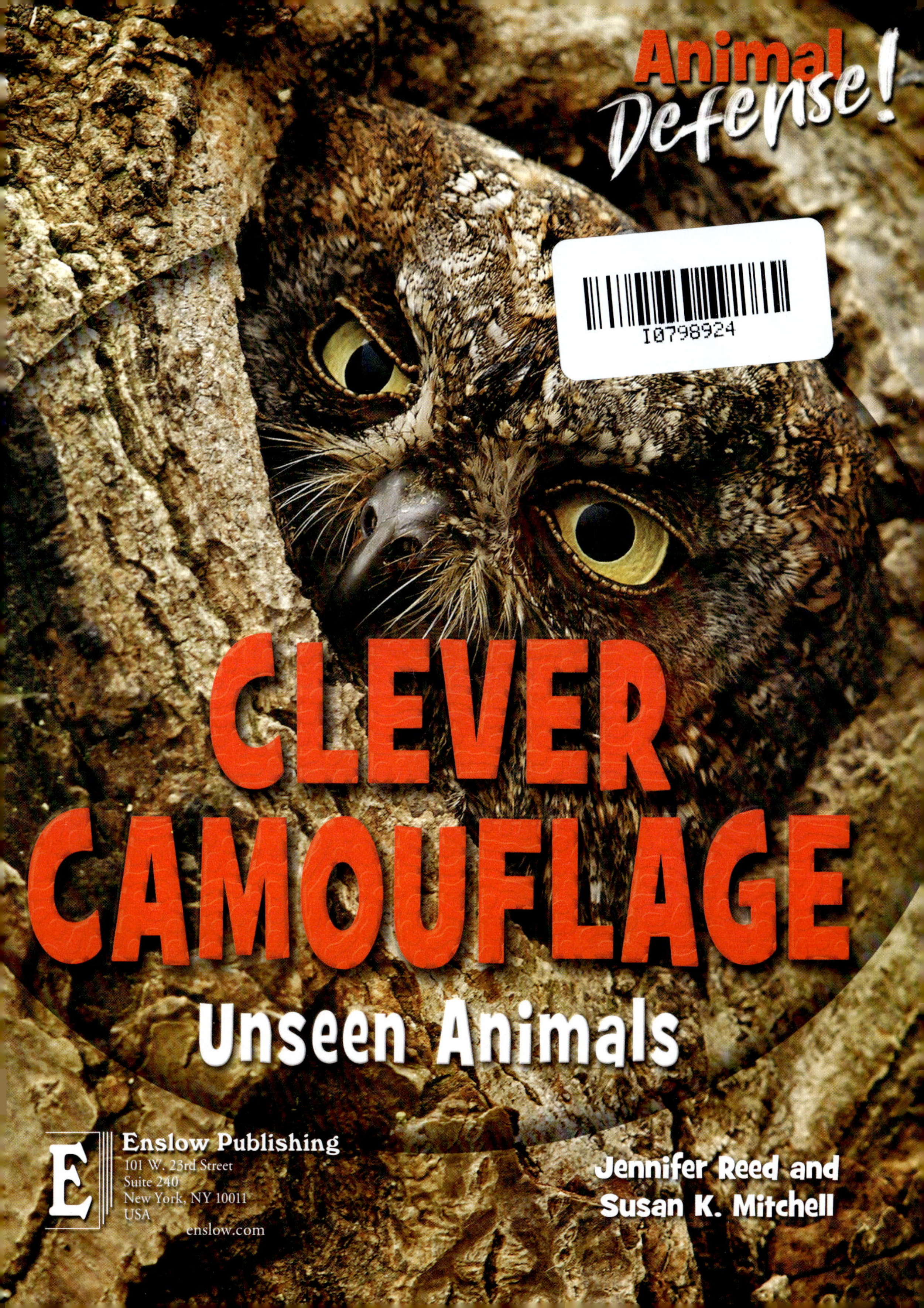

Animal Defense!

CLEVER CAMOUFLAGE

Unseen Animals

Jennifer Reed and
Susan K. Mitchell

Enslow Publishing
101 W. 23rd Street
Suite 240
New York, NY 10011
USA
enslow.com

These books are dedicated to Emily, who inspired the author.
—Susan K. Mitchell

Published in 2020 by Enslow Publishing, LLC.
101 W. 23rd Street, Suite 240, New York, NY 10011

Library of Congress Cataloging-in-Publication Data

Names: Reed, Jennifer, author. | Mitchell, Susan K., author.
Title: Clever camouflage : unseen animals / Jennifer Reed and Susan K. Mitchell.
Description: New York : Enslow Publishing, 2020. | Series: Animal defense! | Audience: Grade 3–6. | Includes bibliographical references and index.
Identifiers: LCCN 2018046197| ISBN 9781978507258 (library bound) | ISBN 9781978508118 (paperback)
Subjects: LCSH: Camouflage (Biology)—Juvenile literature. | Animals—Color—Juvenile literature. | Animal defenses—Juvenile literature.
Classification: LCC QL767 .R38 2020 | DDC 591.47/2--dc23
LC record available at https://lccn.loc.gov/2018046197

Printed in the United States of America

To Our Readers: We have done our best to make sure all website addresses in this book were active and appropriate when we went to press. However, the author and the publisher have no control over and assume no liability for the material available on those websites or on any websites they may link to. Any comments or suggestions can be sent by email to customerservice@enslow.com.

Portions of this book originally appeared in *Animals with Crafty Camouflage: Hiding in Plain Sight.*

Photo Credits: Cover, p. 1 (European Scops owl) David Dohnal/Shutterstock.com; pp. 6–7 Gudkov Andrey/Shutterstock.com; p. 10 Andrew Newman Nature Pictures/ Alamy Stock Photo; p. 13 1001slide/E+/Getty Images; p. 15 Alex Mustard/Nature Picture Library/Getty Images; p. 19 M. van Goor/Shutterstock.com; p. 22 Victor Tyakht/Shutterstock.com; p. 26 Lesleyanne Ryan/Shutterstock.com; p. 29 Muhammad Naaim/Shutterstock.com; p. 32 Scott Camazine/Science Source/Getty Images; p. 33 fntproject/Shutterstock.com; p. 35 Toni Genes/Shutterstock.com; p. 37 Ryan M. Bolton/Shutterstock.com; p. 41 Irina Orlova/Shutterstock.com; p. 43 Vitezslav Malina/ Shutterstock.com; p. 45 Lim Yong Hian/Shutterstock.com.

Contents

Introduction

Animals instinctually need to survive. Over time, they have developed ways to protect themselves from becoming another animal's next meal. They have adapted to their surroundings, making sure to strengthen their defense skills along the way. For some animals, this means that they are able to blend in with their environment so as not to be seen. This makes it harder for predator animals to find and attack their prey. The ability for an animal to blend in with its **habitat** is called cryptic coloration, better known as **camouflage**. The study of camouflage falls under the science of **zoology**. The ability to camouflage well means that an animal can go unnoticed and live another day.

Introduction

Camouflage allows an animal to hide or **disguise** themselves. They become difficult to see. There are four basic types of camouflage: coloration, disruptive coloration, disguise, and **mimicry**. Different animals use different types of camouflage. For example, zebras have black and white stripes. Although zebras don't blend in with their environment, they do blend in with each other. They use disruptive coloration to hide from prey. Zebras often hang out in herds. It is very difficult for a predator like a lion to tell one zebra apart from the group. The colors and patterns on the zebra fit in with other zebras so that there isn't a clear outline of a single zebra. This makes it hard for predators to attack them. Having patterns makes it more challenging for an animal to see another animal, especially from a distance.

No matter where an animal lives, every environment has animals that use camouflage. On land or under the ocean, animals don't only change the color of their skin or fur; some can even change their shape and skin texture, like an octopus. Insects also disguise themselves in all sorts of creative ways. The twig spider is thin and

No two zebras have the same pattern.

green and looks like the stem of a plant. The only time it becomes visible is when it moves.

Humans use camouflage, too. People often wear different clothing—sometimes referred to as camo, short for camouflage—when they go hunting or when they're in the army. Just like animals, their outfits are colored like their environments. Some common patterns match trees, deserts, and water. The colors of camo have become so popular that fashion designers use camo-styled fabric for clothing and interior design.

Take a closer look at specific animals and the different types of camouflage they use in order to survive. Animals use camouflage in the cleverest ways!

Chapter 1

Cryptic Coloration

How an animal will defend or hide itself depends on the environment it's in. Camouflage helps animals blend in with their world. Their fur and feathers act as the clothing they wear. The colors and patterns that many animals have allow them to hide well. The better an animal can blend in with its habitat, the less likely it is to be eaten. Because an animal's environment is always changing, an animal has to be aware of its surroundings and, when needed, modify its modes of camouflage to match the environmental changes. The peppered moth is a good example of an insect adapting its coloring to blend in with the environment. The insect uses coloration and disruptive camouflage to hide from predators. As its habitat changed, so did the coloring on its wings.

Successful Camouflage

Before the nineteenth century, almost all of the peppered moths in England were pale with dark specks. This helped them hide on trees covered in lichens. Lichens are light-colored organisms that grow on the trunks of some trees. Very few peppered moths had dark wings. In this case, the moths used disruptive coloration to hide themselves. It was difficult for their predators to see their outline against the trees, making them hard to spot.

The **Industrial Revolution** changed this. Many factories were built in England's cities. They used coal to power their machines, and the coal made the air dirty. This resulted in fewer lichens growing on trees in the cities, so the trees looked darker.

Scientists noticed that the number of dark-winged peppered moths started growing. The moth was adapting to the darker colored tree and using coloration camouflage to blend in. However, the dark moths were only found near the cities. Out in the country, where

The peppered moth adapts as its environment changes.

the air was cleaner, lichens still grew. This area also had more light-colored moths. Since the air wasn't being polluted, the trees didn't change color. And if the trees didn't change their appearance, neither did the moths.

Changing Skin Color

No matter what color animals are, they all have one thing in common—**pigment** cells. These special cells control the color of skin. In people, these cells make a chemical called **melanin**. The color of a person's skin depends on the amount of melanin he or she has. The more melanin, the darker the skin.

Many animals are covered in fur or feathers. Fur and feathers also have pigment cells. The pigment cells in an animal's skin, along with those in its fur or feathers, help determine what color the animal is. Some animals are born with no pigment. This is called **albinism**. Animals with albinism are white with pink eyes and usually do not survive in the wild.

Fun Fact!

A lion's tan fur matches the light-brown background of the Savannah. This makes it easier for the lion to catch its prey.

Typically, they can't defend themselves by means of camouflage, since they're completely white. Their eyes are also strongly affected by sunlight and, without any pigment to shade their eyes, can be hurt easily.

Chameleons can change the color of their skin quickly. This color change occurs when light is reflected off of their skin and the pigments that are below the surface of their skin are temporarily altered. Chameleons change colors to blend in with their environments. Since their

Red on Yellow Kills a Fellow

Many species use mimicry to camouflage. The harmless scarlet kingsnake copies the colors of the poisonous coral snake to fool predators into thinking it's dangerous. The snakes are both red, yellow, and black. The difference between the two snakes is the pattern of their colors. The scarlet kingsnake is red, black, and then yellow. This rhyme might help: *Red on yellow kills a fellow. Red on black won't hurt Jack.*

Scientists often use albino rats for experimentation.

habitats are often full of green leaves, they are also able to fade in with surrounding trees and bushes. However, it is believed that a chameleon's mood and health are the two most important reasons behind its color changes. A green chameleon may mean that the lizard is relaxed and content.

For many animals, their color makes the perfect disguise. For example, if an animal lives among leaves, being green is very helpful. Brown animals that live around dirt or tree bark can also easily hide. These animals blend into the world around them thanks to their clever camouflage!

Chapter 2

Blending in Underwater

The oceans are filled with amazing creatures. Because the water can be a harsh environment, it is important that the animals that live there know how to protect themselves. Like many animals on land, sea creatures have adapted different forms of camouflage to hide and defend themselves. They also use their abilities to disguise themselves to hunt for food.

The octopus is one of many incredible underwater animals. Few animals can disguise themselves like an octopus can. Disguise is a type of camouflage some animals use for survival. For an octopus, this means changing its skin color and even the texture of its skin. Octopuses use various forms of camouflage, including coloration and background matching. One type of octopus can even take on the shape of its surroundings.

Clever Cephalopod

Octopuses are very shy creatures. Octopuses belong in the same animal group as squids do. They are called

A reef octopus blends in with its surroundings.

cephalopods (SEF-a-low-pods). "Cephalopod" is a Greek word. *Cephalo-* means "head" and *-pod* means "foot." That is pretty much what an octopus looks like—eight long "feet" attached to a head. The "feet" are actually called arms, however.

Inside an octopus's head is a very smart brain. Octopuses are thought to be the smartest of all **invertebrates**, which are animals that have no backbone. Scientists have trained octopuses to do many things. Some have learned to swim through mazes and how to use their arms to open jars of shrimp.

Wriggling under the head of an octopus are eight long arms. Each of an octopus's arms is lined with two rows of suction cups. These are used to catch food and to help the octopus move. The arms can zap out and grab prey as it passes by the octopus's den. They can also feel along the ocean floor for food.

Can You See Me?

An octopus looks like the perfect lunch to many marine animals. Octopuses are a favorite food of some large eels. Dolphins and sharks also like to eat octopuses.

Fun Fact!

The shape-shifting Indonesian mimic octopus is the only octopus that can take on the shape and features of its surroundings.

An octopus does not always have time to get away from a predator. Sometimes there are no nearby places for the octopus to hide. When this happens, the only thing it can do is to try to disappear.

An octopus can blend in with almost any background. Millions of special skin cells help the octopus change colors. Each of these cells can change color independently of the others.

The amount and type of color released determine what color the octopus will turn. How the octopus does this, though, is a bit of a mystery. Do octopuses see their surroundings and change colors to fit in? Not necessarily. Scientists have studied blind octopuses that are able to blend in perfectly with their surroundings.

An octopus can change colors in less than a second. Patterns of color can also form on its skin. This comes in handy since most surfaces are not one solid color.

Camouflage is not the only reason that an octopus turns different colors. Like a chameleon, an octopus's color can show its mood. Researchers have observed octopuses turning red when bothered. Other octopuses have been seen turning white when frightened.

Now You See Me, Now You Don't

Flat fish like the peacock flounder use camouflage to hide from predators. Their skin is similar in color to the ocean floor. They can quickly swim to the ground and nestle into the sand to hide. These creatures need to remain very still. If they move, they risk revealing their true identity and getting eaten!

Blending in Underwater

A peacock flounder matches the sandy ocean floor and becomes invisible to predators.

Octopuses do not only change colors. They can also change the texture of their skin. An octopus's skin is normally smooth. But special muscles in the skin can make the octopus look bumpy like a rock, or grainy like sand. When an octopus changes its texture, it looks a lot like the ocean floor. This helps it blend in even better with its surroundings.

There are a few animals in the ocean that have some very strange camouflage methods. They don't change colors. They don't make their skin look like rocks or sand. Instead, they become practically invisible.

Some shrimp, as well as the glass catfish, are see-through. To other animals, they look like reflected light in the water. Whatever surface the animal lands on shows right through its body. The creature can look like the sand, rocks, or plants around it just by staying still.

Chapter 3

Seasonal Camouflage

Animals live all over the world. In some places, it's hot, and in some places, it's cold. There may be periods full of rain or snow, and there may be times when it's always sunny. Because their habitats change, animals may find that their usual way of camouflaging no longer works. It's important for animals to adopt new ways to protect themselves when the seasons change.

Arctic animals are a great example of this. The Arctic is the name for the far northern areas of the earth. Places such as Greenland, Alaska, and parts of Russia and Canada are considered Arctic regions. The temperature is near or below freezing (32 degrees Fahrenheit) most of the year. Even the summers are quite cold. The yearly average temperature in many Arctic areas is just 7 degrees Fahrenheit (-14 degrees Celsius).

The mammals and birds that live in the Arctic have developed a special type of camouflage. They can change the color of their fur or feathers depending on the season by growing new fur or feathers. This is called **molting**. They shed their winter fur for summer colored fur. Snow camouflage allows animals to turn

During winter, least weasels that live in the Arctic trade their brown fur for a white coat.

primarily white in color, which helps them hide during the winter months. Summer camouflage lets animals change into different shades of browns, gray, and black. The arctic hare and arctic fox change the color of their fur depending on the season. This is seasonal camouflage. Others keep their white fur year round. This is called fixed camouflage.

There are many different kinds of animals living in these icy places. Some are hunters. Some are hunted. They have all learned, however, to survive in the Arctic. One animal that definitely needs to rely on camouflage to survive is the collared lemming.

Snow Survival

Lemmings are tiny mammals called rodents. There are many different types of lemmings. The collared lemming is the only one who can grow a coat of white fur.

The collared lemming's small size makes it an easy victim. If spotted by a predator, the collared lemming will often seek safety under the snow in a burrow.

Fun Fact!

While hunting for seals, polar bears will cover their black nose and mouth with their paws so they appear entirely white.

A hiding place under the snow does not sound very warm, but a lemming's snowy burrow is actually quite comfortable! The collared lemming lines its burrow with grasses and bits of fur. This makes a cozy nest for the lemming. It also provides a safe place to hide.

When they are not resting in their burrows, collared lemmings search for food in the snow. Coming out into the open is dangerous. Unlike some animals, lemmings do not **hibernate** during the winter. They are active all year long, so they are always in danger. They depend on the color of their coat of fur for camouflage.

In with the New

In the summer months, the collared lemming grows grayish-brown fur. It has a lighter stripe of fur around its neck that looks like a collar. This is where the collared lemming gets its name. In the winter, the collared lemming grows new, white fur. It is the perfect camouflage for living in the snow.

Fur, hair, and feathers are not made up of living cells. They cannot change from one color to another the way

Wildlife in White

During the winter, finding food in the Arctic is particularly difficult. Animals who match the icy environment protect themselves from predators and are able to sneak up on their prey more easily. The arctic fox's white fur matches perfectly with the snowy Arctic tundra. These foxes spend a lot of time trailing behind polar bears, eating the scraps left behind after a kill.

The algae on a sloth can also be a yummy snack for them.

that skin can. So, the collared lemming has to grow completely new fur.

As winter gets closer, the days get shorter and colder. There are fewer hours of daylight during the winter. Scientists believe that this tells the lemming's brain to produce certain chemicals. These chemicals then cause the pigment in their hair to change. White fur begins to grow under the gray fur.

By the time there is snow on the ground, the collared lemming is completely white. When spring comes, the days get longer. This time, the collared lemming's fur grows in gray again. This is how the collared lemming is able to camouflage itself all year long.

The Amazon rain forest in South America is nothing like the Arctic. It is extremely wet and hot. Animals in the rain forest, however, face just as much danger as Arctic animals. Things can be especially dangerous for an animal that moves as slowly as the sloth!

The sloth spends most of its life hanging upside down in trees. It has few defenses against predators. A speedy getaway is definitely not possible. The sloth counts on being able to camouflage by blending in with the trees.

The sloth has a naturally brownish-gray coat. In the rainy season, however, a thick green coating of tiny plants called algae grows on the sloth's fur.

It actually makes the sloth look green. Since a sloth moves so slowly, the algae is easily able to grow and thrive. This greenish-colored fur helps the sloth hide in the rain forest trees.

Chapter 4

Masters of Disguise

Insects are able to camouflage in ways that no other type of animal can. Disguise camouflage is when an insect or animal takes on a similar appearance of another natural object, like a leaf, stick, or rock. For example, an insect called the walking stick looks nearly identical to a stick on a tree. There are other insects that also use disguise, like katydids. Katydids are some of the top camouflagers around. Their wings look like leaves. If an insect is living and eating in a tree, then looking like a leaf is the best camouflage there is.

Katydids belong to the same group of insects as crickets and grasshoppers. There are hundreds of kinds of katydids all over the world. In North America alone,

Katydids are poor flyers. Some species can only flutter their wings when leaping.

there are more than two hundred kinds! They come in many colors, shapes, and sizes. But they all have one thing in common—an awesome ability to camouflage.

Leave It to Katydids

Like all insects, the body of a katydid has three parts. There is a head, an abdomen, and a thorax. It also has six legs, just like other insects. But unlike those of many other insects, the large bodies of many katydids are bright green and leaf-shaped.

Fun Fact!

The bird-dropping spider from Australia disguises itself as the one thing birds don't want to eat—their own droppings!

Since being green is not enough to hide the tasty katydid from its predators, its wings actually look like real leaves. Leaves are not completely smooth. Most have veins running down the middle and all over each side. The katydid's wings have the same type of markings. This helps it stay hidden while it eats.

Katydid babies are called nymphs. They look very different from adult katydids. They rely on mimicry to avoid predators. Mimicry is an animal's ability to look like something more dangerous or bad-tasting than it actually is to ward off predators. The animal doesn't hide, but by looking like something unappealing to predators, it is less likely to get eaten. For example, some katydid nymphs look like ants. Many ants deliver a nasty sting, so birds and other insects avoid eating them.

While many katydids look like green leaves, there are also brown katydids. Some have bigger wings than others. Each one has adapted new, individual ways of camouflaging, depending on their environments. For example, the lichen katydid has a lacy, white body.

Colorful Deception

The patterns and colors on a butterfly's wings are beautiful, but they can also be deceiving. The owl butterfly mimics the eyes of owls. A big bold spot on each wing scares away hungry birds.

Monarch butterfly larvae eat milkweed that becomes a poisonous toxin. Birds know to avoid these colorful orange butterflies. The viceroy, another butterfly, mimics the monarch's coloring as a defense mechanism. Birds won't eat a viceroy butterfly if they think it's a bad tasting monarch!

At first glance, the viceroy butterfly (*bottom*) looks like the monarch (*top*). Can you spot any differences?

Its color and shape match perfectly with the spiky, light-colored lichens that it eats. Some katydids are more brown in color which helps them blend in with dead leaves.

Pretty in Pink

Katydids rely on their ability to camouflage to keep them alive. And for most katydids, camouflage works perfectly. Unfortunately, some aren't so lucky. Pink katydids are born with a genetic mutation called **erythrism**. Erythrism affects the color of an animal's fur, feathers, skin, or hair. The mutation can be triggered by the absence of a typical pigment, similar to albinism, or the surplus of the

Pink katydids stand out among the green leaves, making them easy targets.

wrong pigment. For some katydids, this can mean having too much red pigment in their bodies. This is why they look pink.

Sometimes this is actually a good thing for a katydid. If a katydid lives in an environment full of red plants, their pink coloring will help it blend in and be harder for predators to see.

For the majority of pink katydids, though, this genetic mutation only works against them. Since most plants are green, pink katydids are usually unable to camouflage. Without this skill, they have no way to protect themselves. Many pink katydids do not survive to adulthood.

Chapter 5

Coloration Experts

Lizards are part of a group of animals known as reptiles. They rely on the environment to control their body temperature. When a lizard is cold it can be seen sunbathing in an open warm area. Some lizards use camouflage to keep from being eaten. Lizards like the Aegean wall lizard will choose specific rocks that are similar in color to their own skin to warm up on. They blend in nicely with the rocks and are difficult for birds to see. All lizards are experts when it comes to coloration camouflage. They can blend in with their environments whether it's trees, rocks or sand. As long as they don't move, no one knows they are there!

A wall lizard easily blends in with this rock.

Leaf-tailed geckos have some of the best camouflaging skills out of all reptiles. These geckos can easily hide among piles of dead leaves or on a tree trunk. Some have tails that look like brown leaves. The common leaf-tailed gecko of Madagascar is one of the best camouflaged of all the leaf-tailed geckos. Their skin looks like tree bark, a shade of brown that helps it hide incredibly well and go unnoticed.

The Perfect Disguise

Like many geckos, the common leaf-tailed gecko is **nocturnal**. This means that it is active at night. During the day, it sleeps in the jungle trees, flattening itself out on a tree trunk. Its rough skin has splotches of light grays and dark browns. This perfectly disguises the gecko as it snoozes on the bark-covered trees. Its sticky toe pads also come in handy and help hold the gecko on the tree trunk as it rests.

Fun Fact!

When a light-colored gecko is on a dark surface, it takes about an hour to change to a darker color.

Coloration Experts

To make its camouflage even better, the common leaf-tailed gecko has dermal, or skin, flaps. These are floppy flaps of extra skin on the sides of the gecko's body. The gecko's skin flaps seal off any space between the lizard and the tree trunk that might make a shadow that could be seen by a predator. Predators cannot see where the gecko stops and the tree starts!

A mossy leaf-tailed gecko hangs on tightly to a branch that it closely resembles.

Even the gecko's big, stubby tail lies flat against the tree. The tail of the common leaf-tailed gecko is where it stores fat. When there's little to eat, geckos can use the fat stored in their tails as a source of energy.

The common leaf-tailed gecko has more teeth than any other land animal. It can have up to three hundred teeth in its small mouth! But its bite is not as scary as its bark. Geckos are the only lizards that can make noise. They can make little noises like barks, chirps, and squeaks. The common leaf-tailed gecko is the loudest of them all and has a special talent for making a wide variety of noises.

A Grain of Sand

Sand geckos use coloration for camouflage. They often cling to walls and rock crevices in the desert. The top parts of their body, including their eyelids, are a sandy color. They are able to blend in easily with the rocks. Their bellies are white and neutralize the shadow cast by their upper body, making it easier for them to catch food on the desert floor.

When frightened, the common leaf-tailed gecko opens its mouth and shows off a bright red tongue. Next, it makes a very shrill, loud noise. It sounds much like a loud scream. This scream is why some people in Madagascar are terrified of this lizard. It might not be just the scream that scares people, though. Since the common leaf-tailed gecko is so well camouflaged, the scream seems to be coming out of a tree trunk!

Game of Hide-and-Seek

For most animals, life is one big game of hide-and-seek. They spend hours each day hiding to keep from being eaten by a predator. Those few animals lucky enough to have their own camouflage abilities are at the top of that hide-and-seek game. Whether an animal looks like a leaf, or can change the color of its skin, camouflage is truly an amazing animal defense.

Chapter 6

Camo-Wear for Humans

Animals can teach us many things about how to best protect and defend ourselves. Since we can't change the texture of our skin, hide in trees, or make ourselves taste badly, we need to tweak the types of camouflage that we use. Just like all animals, we need to camouflage in ways that are customized for our needs.

Soldiers who wear camouflaged clothes are a great example of human camouflage. Men and women wear military clothing designed to look like their surroundings. These clothes are usually just called "camo."

The patterns and colors of camo have become popular. Fashion and interior designs will often incorporate camouflage into their designs. Non-military people are able to wear and use camouflage styled fabrics every day.

Camo Around the World

Soldiers around the world wore colorful uniforms for many years. Bright-colored coats made it easier to recognize fellow soldiers on a crowded battlefield. It also made it easier to spot the enemy. Early in history,

An ornithologist (bird scientist) camouflages with the woods, making it difficult for birds to see him.

a battlefield could be a very smoky place. Cannons and gunpowder were weapons that produced a lot of smoke.

But as weapons improved, there was less smoke during the heat of battle. Without all that smoke cover, the bright-colored clothes stood out in the greens, browns, or even whites of the land. People in the armed forces realized that being seen during a battle was not a good idea and colorful uniforms were hard to hide. Eventually, they discovered better ways of hiding.

Most armed forces started wearing colors such as browns and dull greens by the twentieth century.

Fun Fact!

Military personnel, hunters, and police use a ghillie suit. It's usually a net or cloth garment covered with leaves, twigs, and twine.

The ghillie suit makes this soldier look like a tree.

These clothes blended in with the land around them. Today, soldiers can choose from many kinds of camo. What they wear depends on where they are.

Tech Savvy Camo

Soldiers who are in a snowy area can wear camo that is white and gray. If they are in a sandy desert area, soldiers can wear light brown and khaki uniforms. In forest areas, soldiers wear dark greens and browns to blend in with the trees. Camo is usually not just one color. It is a pattern of different shades and blotches.

The different types of camo used need to adapt as society changes and evolves. With the advancements of technology, researchers and scientists now have more tools than ever to create new forms of camo. The newest type of camo is called digital camouflage.

Octopus-Inspired Camo

Researchers are working on an electronic camouflage called adaptive camouflage. Adaptive camouflage was inspired by the octopus's ability to change its skin color. It's made from a new material with a temperature-sensitive dye. The material appears black at low temperatures and clear at temperatures above 116 degrees Fahrenheit (46.6 degrees C).

Digital camouflage is created by mixing big and small patterns.

It uses a smaller dotted color pattern instead of large blotches. These small patterns are much like the pixels of a computer screen. There, tiny squares of color work together to create an image. Digital camo uses that same technology to add what looks like texture to color. This camouflage is thought to look more realistic and be better at helping soldiers remain hidden.

Years ago, no one would have believed that this sort of camo could be possible. Who knows what other camouflage strategies, for both animals and humans, the future has in store?

Glossary

adaptive camouflage A material that uses light and temperature to camouflage an object or person.

albinism When an animal is born without any pigmentation.

camouflage When an animal uses its appearance or color to blend in with its surroundings.

disguise When an animal conceals its appearance.

erythrism An inherited disorder that causes an animal's fur, feathers, skin, or hair to be unusually red.

habitat The area in which an animal lives.

hibernate To be in a sleep-like, inactive state.

Industrial Revolution A period of time beginning in the late eighteenth century when machines became more commonly used and many factories were built.

invertebrate An animal without a backbone.

melanin A dark pigment naturally produced by skin cells.

mimicry Having the appearance of a more dangerous or bad-tasting animal in order to fool predators.

molting Growing new fur or feathers with the change of season.

nocturnal Being active at night and at rest during the day.

pigment Color.

zoology The science of studying animals in their habitats.

Further Reading

Books

Harrison, David L., and Giles Laroche. *Now You See Them Now You Don't.* Watertown, MA: Charlesbridge, 2016.

Herrington, Lisa. *Cool Camouflage: Giraffe! Tigers! Seals!* (Rookie Star: Extraordinary Animals). Chicago, IL: Children's Press, 2018.

Levin, Anna. *Hidden in Nature: The Art of Animal Camouflage*. Milano, IT: White Star Publishers, 2017.

Wolfe, Art. *Vanishing Act: The Artistry of Animal Camouflage.* San Francisco, CA: Cameron Books, 2015.

Websites

Explorable

https://explorable.com/camouflage-and-mimicry

Read more about mimicry and how different animals use it for defense.

National Geographic

https://www.nationalgeographic.org/encyclopedia/camouflage/

Explore how different animals use camouflage through text and photos.

San Diego Zoo Animals & Plants

https://animals.sandiegozoo.org/animals/chameleon

Learn more facts about the amazing chameleon.

Index